Making the Best Apple Cider

by Annie Proulx

The heady fragrance of fresh sweet cider running from the press is a wonderful blend of mellow apples, the faintly acidic scent of fallen leaves, and the brisk taste of country air that has been cleared by early morning frost.

Sweet cider is a delicious and refreshing drink, esteemed by centuries of farm workers for its thirst-quenching powers, a tangy beverage savored by children and adults. It can range from a mild apple flavor to a spicy, zesty liquid with a fruity richness reminiscent of ripe pears, to the tingling nip of sharp wild apples. A glass of amber-gold cider poured in winter's gloom and chill has the power to evoke vivid memories of the autumn day it was pressed, and reinforces the sense of thrift and independence that comes from bringing in the harvest and putting up food and drink for the winter.

The memories of family cider-pressing days stay with children for a lifetime, and, through the cooperative work, the fine scents and flavors of the raw material, the instant, gratifying reward of labor in the torrent of delicious golden juice, and the row of filled jugs stored away with anticipation, cider makers enjoy first-hand experiences of self-sufficiency, cooperation, and the pleasure of work.

If you have kids of your own or can borrow a few, let them help with the cider-making — don't hog all the fun.

If you're going to make sweet cider, you might as well make *good* sweet cider. There's too much lackluster, insipid single-variety cider for sale at the "jug mills" that line the road in apple districts, in tourist areas where people come to see the autumn foliage, and in supermarkets. Some of this mediocre cider is squeezed on the premises, some of it is trucked in over hundreds of miles. Some has preservatives added, some is made from reconstituted apple concentrate, but most of the stuff is pressed from the culls of table variety apples — most often McIntosh or Delicious — popular eating apples, but not the best varieties for making either sweet or hard cider.

Even if you've never made cider before, with a little care and selectivity in the apple department, with a few essential pieces of equipment, and with a little help from your family and friends, you can make a better and more flavorful cider yourself at home than most commercial mills ever do. If you've got a discerning palate, a supply of good cider apples, and an interest in fresh apple nectar, you may want to start up your own quality sweet cider business after a few seasons of experimentation.

To make good sweet cider, you need a balanced mixture of apple varieties — some bland for the "base" juice, some tart and acidic to liven the cider up, a few bitter apples with astringent tannins to give body and character, and some aromatic apples for bouquet and outstanding flavor. You will need a grinder to reduce these apples to a pulp, and a press to exert pressure on the pulp so that the fruit cells rupture and release their juice. You need containers to hold the fresh-pressed cider, and, if you're not going to drink it all up within a few weeks, a cold storage place — refrigerator or freezer — is essential lest your sweet cider ferment its way to a potent hard cider.

Equipment
for Cider-Making

Only a few years ago cider-making equipment was scarcer than hens' teeth, but the back-to-the-land movement has motivated an increasing number of manufacturers to turn out butter churns, spinning wheels, horse-drawn farm machinery — and apple grinders and cider presses.

Those nineteenth-century grinder-presses that used to turn up at country auctions have achieved antique status and fetch high prices when they can be found. But an old press in good shape or repaired to usable condition can crank out a lot of apple cider. However, the old grinders with their worn, cast iron rollers can give the juice an unpleasant, metallic flavor, and are rather inefficient at converting hard round apples into soft apple mush. If you use an antique grinder, you may have to run the apple pulp, or pomace, through twice to get it fine enough. In the old days very soft and even decayed apples were often used because they were easier to grind. Rotten apples make rotten cider.

Much of the new equipment is more efficient, easier to clean and lighter in weight than the machines great-grandpa used which were heavy with cast iron and oak. The new grinders feature stainless steel or cast aluminum rollers or breaker teeth, and can be bought ready-made or in kit form. Don't try to put apples through grape or soft fruit crushers — they can't take it.

Small hand and hydraulic presses are available from a growing number of manufacturers, and plans for building your own cider press are available as Report No. 8 from the New York Agricultural Experiment Station, Geneva, N.Y. 14456. This is also the home of the largest orchard of apple varieties in the United States; over a thousand apple varieties are maintained here, many of them true cider apples.

The Grinder: This is usually a hand-powered oak frame set with stainless steel or aluminum cutters or toothed rollers, with a commodious hopper on top which can accommodate up to a bushel of fruit. They are available from several sources, either ready-made or in kit form. Electrically powered grinders vary from the professional hammermill to a small grinder powered by a 3/8″ home electric drill.

The Press: A surprising variety of presses is available to the home cider-maker: small single-tub hand-operated screw presses are ready made and in kits; double-tub screw presses ready made; single- and double-ratchet system tub presses ready made; small hydraulic hand presses with cheese racks which are miniature versions of the large commercial presses; and replicas of nineteenth century presses with some modern modifications.

Pressing Bags and Cloths: If you are using a cheese-rack press, where each rack of apple pomace is enfolded in a nylon press cloth, and the racks are stacked in layers before pressure is exerted, you need press cloths. Usually they come with the press, but replacements can be bought from many sources. If you use a tub press, you'll need a sturdy, porous nylon press bag into which the pomace is poured. They are strong and easy to clean.

Filter Cloth: To filter out stems, seeds, and large chunks of apple pomace from the fresh juice, a layer of cheesecloth or nylon filter cloth is fine. The tastiest and most nutritious cider will be somewhat cloudy with tiny pectin particles in suspension. Filtering sweet cider until it is brilliantly clear is a lot of unnecessary work that takes the desirable pectin out of the cider — purely a cosmetic procedure.

Primary Container: This is the bucket, bowl or vat in which you catch the fresh cider as it pours from the press. It should be stainless steel, odorless polyethylene plastic or nylon, glass or sound, unchipped enamel. Do NOT use galvanized metal containers such as old milk cans, aluminum, copper, or other metal containers, or chipped enamel. The acid in apple juice will react very quickly with the metal and give your cider unpleasant off-flavors.

Plastic Siphon Tubing: A four-foot section of plastic quarter-inch diameter tubing makes a neat efficient job of filling cider jugs and bottles.

Plastic Funnel: For filling five-gallon carboys and other large vessels.

Storage Containers: Earthenware cider jugs stoppered with corks were the traditional cider containers. These can still be found, both old and new, for a price. Most cider makers are thrifty folk who will be content to use clean plastic or glass jugs with screw tops for sweet cider. If you plan on freezing cider, use plastic containers and leave room for expansion under freezing conditions.

SUPPLIERS
OF CIDER EQUIPMENT

Berarducci Brothers
Mgf. Co., Inc.
1532 Lincoln Way
McKeesport, Pa. 15132

Apple grinders; ratchet system hand tub presses.

Day Equipment Corp.
1402 E. Monroe St.
Goshen, Ind. 46526

New and used equipment mostly for the commercial producer. Some small grinders and presses. General supplies.

Which Apples
Make the Best Cider?

"It's the fruit that's most important."

"Nope, it's the cider maker."

Wherever you go in cider country you can expect to hear one or the other lauded as the key to good cider. It's an argument that has gone on for as long as people have pressed apples and drunk the juice.

Hard cider may take the skill of a cellarmaster to make a marriage of the fruit varieties that will weather the rigors of fermentation and grow more beautiful through the flavor changes of aging, but in sweet cider the ideal is an orchard-fresh apple aroma crowning a rich fruit drink. For the best sweet cider you must choose apples whose blended flavors give a juice as piquant and fresh as a bite of ripe apple just off the tree.

Differences in Apples

Apples may appear to differ only in color, shape, and flavor, but for the cider maker there are other important differences. All apples contain the same basic properties: they are made up of 75 to 90 percent water, several sugars (glucose, levulose and sacchrose), malic and other acids, tannin, pectin, starch, albuminoids, oils, ash, nitrogenous substances, and trace elements. Most apples contain roughly the same amount of sugar — between 10 and 14 percent — and though a dessert apple tastes much sweeter to us than a cooking apple or a tart wild apple, it's not because the dessert apple contains more sugar, but because higher levels of malic acid in "sour" apples mask their sweetness. A good rule of thumb is that a sweet-tasting apple is low in malic acid.

The tannin content, acid levels, and aromatic oils determine the value of the different varieties for cider makers. Tannin causes an astringent puckering sensation in the mouth, and malic acid contributes a tingling sourness. They give body and zest to a hard cider, and, in moderate proportions, contribute to the perky freshness of a good sweet cider.

In Europe special cider apples, sometimes‛ called "vintage" apples, are grown. These are small apples with tough, russeted skins and have a sharp, bitter taste because of their high levels of tannin and acid. They are used exclusively to make hard cider, and are unacceptable either for table fruit or fresh juice.

In North America most cider is made from culled or surplus dessert apples. The sugar and acid levels in these apples are good for both hard and sweet cider, but there are few aromatic types, and most dessert apples are sadly deficient in tannin, so are not ideal for cider. The cider maker can overcome this latter deficiency by blending in wild apples or crab apples. Both have good levels of tannin.

'Perfect' Cider Apples

There are "perfect" cider apples, those varieties which have good balance of aromatic oils, sugar, tannin, and acid, but they're rare and becoming rarer. The Roxbury Russet, Golden Russet, Ribston Pippin, and the Nonpareil were long treasured as fine single-variety cider apples. These antique varieties are hard to find today, so most cider makers have to make do with the varieties at hand, and balance their cider through blending. Most of the best ciders and apple juices of the world are blends, and the better for it.

Blending

Rather than "shotgun" blending, where as many different varieties as are available are fed into the press in one giant cider gamble, a system of judicious blending gives a more predictable, superior product. Popular varieties can be grouped by their cider characteristics, and the trick is to get your hands on the types needed to round out a blend.

There are thousands of different apple varieties, many of them members of the same apple family with strong similar characteristics. The McIntosh group is popular and widespread, and several of them unwittingly blended in a cider could produce a batch of juice overpoweringly scented with McIntosh perfume. Macoun,

Cortland, Spartan, Jonamac, and Empire are all McIntosh descendents. In the same family, but further removed and so better candidates for blending with the very aromatic McIntosh are Quinte, Ranger and Caravel.

"Wild" apples make better hard cider than "tame" apples — the cultivated varieties —, but in sweet cider a straight or high percentage of wild apple juice in a blend can make a cider that is too bitter and sharp for tastes acclimated to sweet juice drinks. Hard cider made from wild apples is more palatable than the fresh juice because the second stage malolactic fermentation neutralizes most of the malic acid. However, if wild apples are available, and your cider base is bland and lacking in character, you can blend in wild apple juice or crab apple juice as you would the juice of domestic varieties with higher acid and tannin levels.

'Wild' Apples

What are "wild" apples? They are not the fruits found on old domesticated cultivars standing beside cellar holes or in neglected, long-abandoned orchards. Although disease, vermin, and poor nutrition can make the fruits of these trees look like their wild cousins, they are still of proper family no matter how scruffy, and the apples will have the family characteristics. Wild apples are the naturally seeded offspring of known cultivars, or are from trees which have grown up from rootstock suckers bearing fruit of an unknown type and quality. (Not all wild apples are good cider material; some are mushy and characterless. Taste samples before you gather them, and if they are bland and insipid, pass them by.) Since apples do not breed true to seed, but are propagated by grafting, the seeds of the familiar orchard varieties which escape captivity and grow to maturity tend to bear fruit which resemble those of their ancient ancestors, the crab apples.

Some Things to Think About

Apples unacceptable for the market and surplus apples have long provided the bulk of cider fruit in North America. This may be a good way to utilize a perishable commodity, but unfortunately the cider press has also been used to clean up the orchard of battered windfalls, natural drops, and even decaying fruits. The best ingredients make the best products, so to avoid the plague of off-flavored or vinegary cider, use only sound, ripe fruit. Immature apples have low sugar levels and are too acid and starchy for cider, while drops which have been on the ground for a while can taste moldy. Brown-spotted and rotten apples are loaded with vinegar-inducing acetobacter. Windfalls can also contain a toxic substance

A GUIDE TO CIDER CLASSIFICATIONS

High Acid	Medium Acid	Low Acid
Close	Baldwin	(Neutral Base)
Cox's Orange Pippin	Cortland	Ben Davis
Esopus Spitzenberg	Empire	Delicious
Gravenstein	Fameuse	Golden Delicious
Jonathan	Golden Russet	Grimes Golden
Melba	Idared	Lindel
Newtown	Jerseymac	Rome Beauty
Northern Spy	Lobo	Westfield Seek-No-Further
Quinte	McIntosh	
Rhode Island	Rambo	
Greening	Roxbury Russet	
Ribston Pippin	Sops of Wine	
Vista Belle	Spartan	
Wealthy	Wayne	
	Winesap	
	York Imperial	

Aromatic	Astringent (Tannin)	
Cox's Orange Pippin	Dolgo Crab	
Delicious	Geneva Crab	
Fameuse	Lindel	
Golden Delicious	Mont Royal	
Golden Russet	Newtown	
Gravenstein	Red Astrachan	
McIntosh	Siberian crab apples	
Ribston Pippin	Most wild apples	
Roxbury Russet	Young American crab	
Wealthy	apple	
Winter Banana		

known as patulin. Both unripe apples and aging drops make very poor cider, sweet or hard.

How to tell when an apple is ripe? Twist it clockwise on the stem. If the apple comes off the tree readily, it is usually ripe. A further test is to cut the apple open and look at the seeds. If they are dark brown, the apple is mature. Light tan or pale white seeds mean the fruit is not ready for harvesting. Sometimes in late summer or September high winds will knock down many maturing fruits which are not quite ripe. Do not be tempted to use these for cider. Instead, use them for apple or herb-flavored jelly.

McIntosh, Winesap, and Delicious apples should never be used to make a single variety cider. Winesap has an unpleasant, bitter flavor by itself, and Delicious, which is very low in acid, makes a bland, insipid cider. It does have good sugar levels and pleasant aromatic oils, and can be used as a suitable base with more lively varieties such as Jonathan or Newtown blended in. McIntosh has a uniquely penetrating aroma which drowns out other desirable cider characteristics. Use it in blends if you like the McIntosh essence, as some people do.

In the final analysis, blending is a process of taste trial and error until you hit on a combination of varieties that fits your palate. This will be your unique, personal cider, unlike any other.

Finding Apples

If you do not have apple trees of your own, you may want to plant several varieties of fast-maturing semi-dwarf trees strictly for cider production. If you decide to gather wild apples from the wayside or from an abandoned orchard, be sure to get permission from the owner of the land; he or she may be planning on using them for cider too. Experimental Station orchards often have hard-to-find varieties for sale to the public. Commercial orchards and pick-your-own orchards are other apple suppliers.

If your cider fruit comes from an orchard that has been heavily sprayed, you may want to wash the apples, then cut off the stem and blossom ends before milling them.

Steps to Making Cider

The flavor and quality of your cider will depend not only on the varieties and proportions of apples you use in your cider blend, but also on the summer weather as the apples ripened, on orchard care, on mellowing techniques, and your own personal taste. Make cider outside, preferably on a cool, breezy day. The low temperature will reduce the risk of bacteria growth, and the breeze will keep away the tiny vinegar flies that can infect your new cider with *acetobacter*, the bacteria that make vinegar — and disagreeable sour cider.

An extremely important part of quality control in cider making is cleanliness. All materials and equipment must be clean and sanitary. The press, grinder, and all utensils should be scrubbed and hosed down with plenty of clean water (no soap) after each day's run is over, even if you plan on pressing again the next morning. If your equipment is in a cider house, the walls and floors must be hosed down at the end of the run also, to prevent the growth of acetobacter and to avoid providing pleasant surroundings for the dreaded vinegar flies. Dirty cider equipment leads directly to ruined cider.

1 **Apple harvest and "sweating":** Harvest or buy only mature, ripe, sound apples. Do not use unripe apples or windfalls. Immature apples make inferior cider.Windfalls are loaded with undesirable bacteria which will contribute unpleasant off-flavors to the juice.

Store the harvested apples in a clean, odor-free area for a few days to several weeks, "sweating" them until they yield slightly to the pressure of a firm squeeze. This mellowing procedure improves the flavor of the cider and makes the apples yield up their juices more readily. Never use rotten or decaying apples. Keep the different varieties separate if you want to make a balanced blend after pressing.

Be extra careful washing the apples if they have been sprayed.

2 **Selecting apples for blending:** You can expect to get about three gallons of cider from every bushel of fruit. A bushel of apples weighs forty-five pounds. Depending on the varieties you are able to gather, you can make a good blend based on the following amounts of fruit in each category. Check the lists on page 9 for the varieties.

Neutral or low acid base: 40 to 60 percent of the total cider. This bland, "sweet" juice will merge and blend happily with sharper and more aromatic apples.

Medium to high acid: These tart apples can make up 10 to 20 percent of the cider. Use any of the listed varieties which please your palate.

Aromatic: 10 to 20 percent of these fragrant apples will give your cider bouquet. Most of the aromatic oils are concentrated in the apple skins, but these cells are difficult to rupture. An efficient grinder helps.

Astringent (tannin): 5 to 20 percent of the total juice. Go easy with these tongue-roughening varieties. Too many will give your cider a fierce puckering action on the taste buds. The North American varieties which are high in tannin are also high in acid, so blend them carefully.

3 **Washing:** Just before grinding, wash the apples in a large tub of clear, cool water. Squirt a garden hose at high pressure directly on them. This helps get rid of unwanted bacteria and orchard detritus. Toss out any rotten or moldy fruit.

4 **Milling or grinding:** Dump the washed apples into the hopper of the grinder and reduce the fruit to a fine, mushy pomace. In ancient times apples were pulped for cider in stone troughs by men swinging heavy nail-studded clubs called "beetles." North American dessert apples are quite soft and make a pulp that may be even too mushy for efficient pressing. Wild apples give a more manageable pomace. Very small batches of apples can be mushed in a food mill or chopper, but this is impractical if you want to make more than a gallon or two of cider.

If this is your first cider-making experience, keep the apple varieties, the ground pomace, and the juice separate and blend to taste later. When you have worked out a desired mixture with known quantities of varieties, next time you can grind the mixed

Grind the apples to a pulp.

apples all together. Many cider makers enjoy experimenting with new blend ratios every fall. That's what makes good, individual ciders. Do not let the pomace stand, but press it immediately.

5 **Pressing:** If you are using a single-tub screw press or ratchet press, place the nylon press bag in the tub and fill it with apple pomace. Do not use galvanized or metal scoops which react with the acid in the pomace. Tie or fold the bag closed, and slowly apply increasing pressure to the pulp. Don't hurry the process — you may burst the bag of pomace. As the juice flows out you can tighten the screw or pump the ratchet to bring the pressure up again. It will take about twenty-five minutes for the juice to come out, no matter what kind of press you use.

If you have a double-tub press, you can set up a continuous cider-making operation with one or two helpers. As one person grinds the apples which fall in pomace form down into the rear tub beneath the grinder, the other tightens the screw or ratchet which presses the pomace in the front tub. When the juice has all been extracted, the front tub slides forward out of the press, and the pomace is dumped. The back tub, now full of fresh-ground pomace, slides forward to be pressed, while the just-emptied tub is

Place the pressing disc in position.

...ess the juice from the pulp.

popped under the grinder in the rear of the press for a new load of pomace. This "continuous" method yields about eight to ten gallons of cider an hour, and is good exercise for whomever twirls the grinder handle.

The small hydraulic presses are designed to press pomace which has been loaded into a number of slotted racks, each lined with a sturdy but porous nylon press cloth. The pomace is dumped onto a press cloth laid over a rack. Then the cloth is folded over on each side to completely enclose the pomace, and another rack placed on top of it. The process is repeated until six to eight layers or "cheeses" are stacked up. The hydraulic press, usually operated by a foot pedal, then exerts pressure, and the press stands until all the juice has flowed from the pomace, about half an hour. This is the type of press, in large size, most often seen at commercial cider mills.

Catch the juice in a container.

Catch the fresh cider in a clean stainless steel or plastic or sound enamel container. Do not allow your cider to come in contact with other metals. If you are planning to blend the juices from different varieties rather than mixing predetermined ratios of apples before grinding, keep the pressed-out juices separate, but covered and cool until you have a chance to blend. Put the pressed-out pomace to one side. If you are going to use the pomace to make the light, refreshing drink called "ciderkin," dump it into a clean plastic garbage can as you go along, and keep it covered. Directions for making ciderkin are on page 27.

Custom pressing at the cider mill: If you don't have a cider press and do have a lot of apples, find a cider mill that will do custom pressing. Phone ahead for an appointment, for in the cider season most mills operate long hours to keep up with the harvest. Some mills will charge you a per-gallon rate for pressing your apples. This can be expensive — as much as seventy-five cents per gallon. Other mills will barter their pressing services for your extra apples. Yet other mills will take *your* apples and give you *their* juice. If you are aiming for a high quality, personally blended product, avoid this last mill, for you have no control over what apples went into the mill's pressing.

Most mills refuse to press less than ten bushels of fruit. Before you show up at the mill be sure they'll take your apple varieties for pressing. Many mills refuse to take crab apples that jam up their grinder machinery because of their small size and hard consistency. A few mills with high standards refuse to press soft dessert apples on the grounds they make poor cider. No self-respecting mill will even look at dirty, half-rotten fruit which will contaminate their machinery. A very few mills have small, hand-operated grinders and presses on their premises for the use of customers who only have a bushel or two of apples to press. Take non-metallic containers to transport your juice home. A clean, sound barrel is convenient. Be sure to wash it out thoroughly after you siphon the fresh juice into smaller containers.

6 **Blending the juices:** If this is your first cider, it will be helpful in future autumns to know the proportions of different apple varieties you used to make the delicious beverage. A few scribbles on the spot in a small notebook can save you time and headaches next year. Serious cider drinkers will eventually invest in a hydrometer to measure the sugar content of the juice, and an acid tester for determining acid levels. Others prefer to blend strictly by judicious tasting.

Take quart samples of each kind of juice. Use a measuring cup to figure exact amounts, and try for a good balance of juices. Taste-test for tannin content first. Add small amounts of high-tannin cider to the neutral or low-acid cider base until the level of astringency is pleasing to you. Look for a sensation of dryness or puckering in your mouth when you are trying to isolate a tannin taste reaction from an acid reaction. A lot of tannin in a juice will make your tongue feel rough, as though each taste-bud were standing on end. A lot of acid in a juice tastes sour and sharp, but without that dry, puckery sensation. Next, add in aromatic juice, then cautiously blend the high acid juice until the cider is lively, fragrant and well-balanced. Referring to your notes, blend your bulk juice to the same proportions. The cider is now ready for filtering and storage.

Experimental Station-tested balances of juice types for tasty sweet cider can be your primary blending guide. The total cider consists of ten equal parts.

And enjoy.

3 to 6 parts: low acid or neutral juice for the cider "base"
.5 to 2 parts: high tannin juice
1 to 2 parts: aromatic apple juice
1 to 2 parts: medium to high acid juice

Ratios can vary considerably, depending on personal taste.

7 **Filtering:** Not so many years ago Americans preferred glassy clear cider with the pectin content removed by filters and enzymes. Current preferences lean toward natural, unadulterated, unrefined foods, including the faintly hazy, natural sweet cider which has passed through only a light layer of cheesecloth or nylon mesh to catch impurities and flecks of pomace. The pectin and crude fiber in a natural sweet cider supplies bulk in the diet and regulates the digestive system. Pectin has also been linked in recent experiments with regulating cholesterol levels in humans. Most nutritionists now recommend sweet cider with its natural pectin content as superior to the crystal-clear type.

Storing and Preserving
Sweet Cider

Refrigeration: Cider can be stored for short periods in clean glass or plastic jugs, or in waxed cardboard containers in a refrigerator at normal refrigeration temperatures. Depending on the condition of the fruit used to make the cider and the cleanliness of the grinding and pressing machinery, it can stay fresh-tasting from two to four weeks. If the machinery was not properly cleaned and harbored substantial colonies of yeasts and bacteria, you can expect fermentation to begin soon, or undesirable bacteria may grow.

Freezing: After refrigeration, the preferred method of preserving cider is freezing it in plastic or waxed cardboard containers. Most families use the half-gallon size as the contents can be consumed in a couple of days, but if large quantities of cider are at hand, this is a bulky procedure. Allow two inches in the necks of the containers for expansion during freezing. Defrost the cider for a day in the refrigerator when you want to drink it. You can keep frozen cider for a year with little deterioration in quality.

Pasteurizing in Bulk: Cider can be stored almost indefinitely on the pantry shelf by pasteurizing the juice and then preserving it in bottles or canning jars. This is the hot pack method using sterilized bottles with crown caps or regulation canning jars with new lids. If you use bottles and caps, you will need a bottle capper to affix the caps. You also need a metal-stemmed high temperature thermometer, such as the kind used for candy-making or deep frying. You will need a large kettle to sterilize the containers and keep them hot, and a stainless steel or unchipped enamel kettle in which to heat the cider. The acids in the juice will react with other metals and taint the cider. Bring the cider up to a temperature of 160° F. and hold it there for fifteen seconds. Fill and cap the containers, place them on their sides on several layers of paper in a draft-free place, allow them to cool, then check to be sure they are properly sealed. Store in a cool, dark place.

Pasteurizing in the Bottle: Another method of pasteurization — one that retains more apple flavor because lower temperatures are used — is the *Holding System* where bottles are filled with cider, capped, and then pasteurized. Here you'll need cappable bottles of one quart size or smaller, a water bath canner, and a high temperature thermometer. Save one bottle for a control bottle: into the neck of this bottle fit a cork which has been drilled through to snugly accept the thermometer stem, after filling the bottle with water. Fill the other bottles with cider, and cap. Place them in the canner rack and fill the canner with water until all the bottles are completely covered. Put an upside-down clean tuna fish can at the bottom of the canner and on it stand the water-filled control bottle. The cork and thermometer should be above water. Now heat the water until the thermometer registers 143° F. Maintain this temperature for thirty minutes to complete pasteurization. Remove the bottles and let them cool on their sides away from drafts on several layers of paper. Store the cider in an upright position in a cool, dark place. If pectin and vegetable matter settle at the bottom, shake the bottle before serving.

Cider Concentrate or "Boiled Cider": If freezer storage space is a problem — and there's nothing like twenty or thirty gallon jugs to make a problem — you may want to concentrate that sea of cider.

The best quality Cider Concentrate is made on a cool day in the 40's by pressing the apple pomace as soon as the apples are ground, and by hustling the fresh pressed juice into your evaporating pan, a shallow stainless steel or enamel pan. Cider that has been allowed to stand several hours will not make good concentrate.

Reduce five gallons of very fresh cider to one gallon over a brisk fire. The process should be as rapid as possible to capture the full flavors. The concentrate is done when the temperature reaches about 216° F. Strain the liquid through a double layer of cheesecloth, cover it carefully and let it stand twenty-four hours until any sediments sink to the bottom. Siphon the heavy liquid off into hot sterilized canning jars, and process them in a boiling water bath for fifteen minutes. Cool them in a draft-free place and store in cool darkness. This tasty concentrate can be reconstituted with cold water — four or five parts water to one part concentrate. It can also be added to mulled cider drinks, to a pot of baking beans, or to season the best apple butter, apple sauce, and mincemeat.

Hard Cider

Making fine hard cider can be as complex a process as making a fine wine. Scores of factors, ranging from the pH level of the juice and the variety and age of the apple trees to the kinds and properties of yeasts and fermentation temperatures and procedures, can radically affect the flavor and quality of the finished cider. Only the basic steps to making a simple farm cider are covered here, but this is the cider that sent many a plowboy capering up the cellar stairs as though his chore boots were ballet slippers. A natural hard cider to which no sugar has been added, can attain a respectable 6 or 7 percent alcohol if fermented to dryness.

What is Hard Cider?

Hard cider begins where sweet cider leaves off. The fresh-pressed apple juice (called "must" by hard cider makers) is poured into a fermentation vessel. The fermentation which follows is often in two stages. In the first phase, the natural yeast flora in the apple juice feed on and convert the natural hexose apple sugars into alcohol. When the sugars are exhausted, alcohol production ceases. At this point a secondary fermentation may take place, either in the fermentation vessel or the storage bottles. This is the fermentation of the malic acid in the cider by lactic acid bacteria. The result is lactic acid, a more mellow and subtle-tasting acid than the familiar sharp malic acid of fresh apple juice, and carbon dioxide bubbles.

Barrels — Traditional but Troublesome

Cider was traditionally fermented in barrels, and many people still believe that good hard cider needs an oak barrel. However, barrels are temperamental, difficult vessels now just about obsolete in commercial hard cider operations; preferred are high density polyethylene tanks and glass-lined steel vats.

Availability and tradition may make barrels your choice, but here are some common barrel problems. The large cellular pores of wooden barrels absorb off-flavors easily, hold harmful bacteria, and expand and contract. If you're thinking of buying a second-hand barrel for cider fermentation, smell it carefully. If it smells of vinegar, mold, mustiness or is just plain nasty, *do not use it* for hard cider. More hard cider is ruined by barrel contamination than any other cause. Barrels which have contained *any* liquid in the past except cider, whiskey, applejack, sherry, brandy, rum, or port should never be used for cider fermentation. Old cider barrels may not have been cleaned thoroughly after use, and may harbor millions of acetobacter which will remorselessly convert your cider to vinegar or a very common ghastly beverage which falls somewhere between cider and vinegar. Old whiskey barrels and other hard liquor barrels which have been used *more than once,* will have absorbed unpleasantly high levels of fusel oils and alcohol. The fusel oils will give your cider off flavors, and can give you a roaring headache if you drink it. The alcohol residue can actually inhibit the growth of the yeast and the proper fermentation of your cider.

A common mistake made by the hopeful cider maker is to pour a fifth of whiskey or rum into the fresh cider barrel in hopes of making the cider "stronger." The alcohol will slow down or stop the fermentation process. If you want to add liquor to your cider, do it in the glass after the fermentation process is complete.

New barrels can give cider a strong and unpleasant "woody" taste unless they have been carefully treated and prepared. Both old and new barrels should be steam cleaned, rinsed, sterilized, and rinsed again before you pour the apple juice in. This process must be repeated when the barrel is emptied at bottling-off time, to prevent any residue from going acetic. Then the barrel must be topped up with fresh sweet water, rotated every few weeks, and the water changed every two months until the barrel is used again. Barrels must never be stored empty or allowed to dry out.

If you still want to use barrels for fermenting cider, remember that they must never come in contact with the ground or cellar floor, but should lie in a "barrel cradle," and must be kept topped off — filled to the brim with additional cider — during fermentation to prevent the barrel top from drying out and allowing bacteria-laden air to come in contact with the maturing fluid. The best barrels are sound, clean whiskey, rum, or sherry barrels which have been used *only once* before. Most garages can and will steam clean barrels for you. Special barrel sterilizing compounds are available from wine supply houses.

If you are making hard cider for the first time, your chances for a good-flavored cider will be vastly improved if you use five-gallon glass carboys or polyethylene containers for fermenting. Glass carboys are sometimes available from bottled water suppliers, and the less expensive poly containers are found everywhere at discount and hardware stores. They are cheap, light, unbreakable, easy to clean, and take screw-on, air-tight fermentation locks. Both can be purchased from wine supply stores. Two mail-order wine and cider-making equipment suppliers are:

Vynox Industries, Inc.
400 Avis St.
Rochester, N.Y. 14615

Happy Valley Ranch
Rt 2, Box 83
Paola, KS

Making Basic Hard Cider

The single cardinal rule to making hard cider is *never let air come in contact with the cider at any stage of fermentation or storage.*

1 **Blend the fresh-pressed sweet cider** immediately after pressing. Do not allow the juice to stand exposed to the air more than necessary. Exposure to air increases the chance of contamination by acetobacter which will convert your cider to vinegar.

2 **Strain the blended must** into your fermentation vessel right to the top in a cool, clean area. A clean, unheated cellar is ideal. Don't make cider in any place where strong-smelling substances such as oil storage tanks, gasoline, rubber, cleaning materials, onions, or root crops are stored. The cider will readily absorb such odors and have peculiar and unpleasant off-flavors. The higher the temperature, the faster the cider will ferment, but a slow fermentation rate makes a better quality cider. A good temperature range is 50° F. to 60° F. though decent ciders have been made in warmer areas.

Set aside a gallon of fresh sweet cider, covered, in the refrigerator for topping off the working cider. Leave the top of your fermenting vessel open. The liquid will not be contaminated by air at this stage because the cider is giving off gas. Within a few days the cider will be visibly "working" and will boil over a thick foam as it casts off impurities and detritus in the first vigorous stage of fermentation. Keep the foam wiped off the sides of the container. Add a little fresh cider daily to keep the liquid level high, for as long as the cider is working and the level is up to the top of the opening, the chances of air-borne acetobacter contamination are nil.

As the foaming subsides, screw on a water fermentation lock and allow the cider to ferment at its own pace. This process can take from a few weeks to five or six months, depending on the temperature of the cellar, the amount of natural sugar in the juice, the vigor of the natural yeasts and the chemistry of the apples. When the bubbling in the waterlock subsides, the yeasts will have exhausted the sugar in the cider; the cider has now fermented to dryness and is drinkable.

3 **Siphon the finished cider** with plastic tubing into sterilized bottles, and cap them with crown caps, available from wine supply stores, or screw-on caps or corks. Store the bottles on their sides if you use corks so that the liquid will keep the corks swollen and tight-fitting. A cool, dark corner of the cellar is a good place to store the precious golden fluid, now your own natural hard cider. You may drink it right away, or years from now.

If you are making cider for the first time with a new press, the natural yeasts on the apples may not build up in sufficient quantity to give you a good, vigorous fermentation. Many cider makers find it helpful to guarantee a good ferment by adding a commercial champagne or white wine yeast to the must when it is poured into the fermentation vessel.

Cider makers who want a more potent drink, or who suspect that the natural sugars in the fresh apple juice are low because of a cold, rainy summer, may want to add sugar to the must for a higher alcohol content. A hard cider with an alcohol content of less than 5.7 percent does not keep well. By adding about one cup of sugar per gallon of must before fermentation, your finished cider will have roughly 10 to 11 percent alcohol. Adding too much sugar can stop the fermentation process.

Old New Englanders often added raisins instead of sugar; a few handfuls of natural raisins to a five-gallon jug of juice supply the fermentation process not only with additional sugars, but with wine-type yeasts from the raisin skins.

Making hard cider without a few instruments to measure sugar and alcohol levels is a seat-of-the-pants procedure. Some of this b'guess-and-b'gosh cider can be very good, depending on the expertise of the maker, the weather, the apples, and the luck of the game.

The Beefsteak Myth

Somewhere along the line word has gotten out that hard cider needs a big juicy beefsteak tossed into it to "ripen it up." What happens most often when this is done, is putrefaction, not fermentation, resulting in a foul liquid that not even the devil could drink. In the old days, a fermenting cider sometimes stopped bubbling in the midst of the process (known as a "stuck fermentation") and the desperate cider maker, not realizing that the problem was a lack of nitrogen in the juice, knew only that adding a piece of meat to the barrel would start the fermentation process up again. The decomposing meat added enough nitrogen to the cider to boost the fermentation. The people who drank this cider, with its rank off-flavor, called it "scrumpy." Today, "nutrient tablets" sold by wine supply houses will restart a stuck fermentation without recourse to the meat counter.

Vinegar

From the giant acetator towers of commercial vinegar factories to the forgotten gallon cider jug in the pantry, vinegar is made many ways. Though a lot of vinegar has been made by accident, the best is made with care.

Vinegar is the result of a secondary fermentation of hard cider. Acetic bacteria, or acetobacter, in the presence of oxygen will change the alcohol in hard cider to acetic acid. The bacteria is either airborne, transported directly to the cider on the feet of vinegar flies attracted by the yeasts, or deliberately introduced in a culture. Since it is the *alcohol* which converts to acetic acid, leaving a jar of sweet cider open to the air and hoping for good vinegar to develop is a chancey undertaking. The yeasts in the cider must convert the sugars to alcohol before acetic acid can develop. Since many bacteria and molds in the air feed on the fruit sugars in sweet or partially fermented cider, there is a considerable risk of making a foul moldy liquid instead of a sharp zesty cider vinegar, unless you start out with a hard cider which has been fermented to dryness.

To make good vinegar, first make hard cider, as explained earlier. (Before going on to the next stage, remember never to make vinegar anywhere near sweet, fermenting, or stored cider; the risk of acetobacter contamination is high.) Pour the hard cider into a wide-mouthed container — glass, glazed pottery, stainless steel, or enamel — for the maximum exposure to oxygen and acetic bacteria. Cover the container with several layers of cheesecloth to keep out insects or mice. Don't worry about the acetobacter — they'll pass right through the cloth. The vinegar crock should be left in a fairly warm place in dim light. Sunlight has an inhibiting effect on the development of the vinegar.

The conversion of the alcohol in the cider to vinegar can take several weeks or several months. If you have made a natural hard cider without adding sugar, it should contain about 6 percent alcohol. This will convert into an equal amount of acetic acid. A 6 percent acetic acid vinegar is a good table strength. Anything more, acid will have to be diluted with distilled water, for it will be extremely sharp to taste.

Mother of Vinegar: When the cider has changed to vinegar, you will notice the gelantinous mass of acetobacter floating in the crock. This is the famous "Mother of Vinegar" or "vinegar mother" and it has value. A good vinegar mother can be sold, bought, or traded, for it serves as a catalyst which speeds up the conversion of your hard cider to vinegar. Vinegar makers in a hurry can buy vinegar mother at many health food stores.

Bottling: Full strength vinegar made from completely fermented hard cider can be simply poured into sterile bottles, capped, and stored. If you make vinegar from incompletely fermented cider which still contains sugar, you must pasteurize the vinegar in a hot water bath if you want it to keep for any length of time. When capping vinegar bottles use coated lids, cork-lined crown caps, rubber and glass covers or just corks. Metal lids must be coated, as the acetic acid in vinegar is highly corrosive to metals.

What to Do
with Left-Over Pomace

After the pressing is over you will face a mountain of damp pomace — the skins, cores, seeds, stems and pulp which has been wrung free of juice. Many people view this mass of browning pomace with dread — a waste product which has to be gotten rid of somehow. But pomace has many uses, and if you can't do something with it yourself, try swapping it with someone who can. Farmers and sheep raisers are delighted to have pomace to feed to their stock. Here are some other uses.

Ciderkin and Mock Cider: Ciderkin was a popular children's drink in the old days, and it's still refreshing and tasty to modern palates. It is made by soaking the still aromatic pomace which contains plenty of sugars, yeasts, and flavor, in water overnight. Then run the reconstituted pulp through the press again. The result is a delicate, sweet cider drink which can be enjoyed straight from the press. If sugar and yeast are added to this juice it can be fermented into a mock cider.

Feed for Livestock: One part of pomace blended in with four parts of cattle or sheep feed makes an excellent animal food. Stock relishes pomace, though feeding it straight in large amounts can cause diarrhea. Many sheep raisers like to finish lambs for the fall market by feeding them apple pomace for several weeks. Pigs also enjoy pomace. If you don't raise animals yourself, try swapping or selling your pomace to someone who does.

Seedling Stock: Did you ever wonder where nurserymen get all those sturdy rootstocks used for grafting cultivars onto? Most of them are the strong seedlings sprouted from the seeds in the pomace of commercial cider mills. If your pomace hasn't been smashed up in a hammermill, you can start your own apple-crab nursery by spreading some of the seediest pomace in a freshly turned field, or by starting individual seeds in an apple seedling bed. When the trees are several years old they can be used as rootstock for your favorite budded or cleft-grafted cultivars. You can let these trees grow to maturity, too. This will take a long time, but who knows, one of them might turn out to be the greatest cider apple of all time. It's a gamble.

Compost: Pomace is so acid it must be composted for two years before it can be applied to the garden. By that time composting temperatures will have sterilized the seeds, or you'll have pulled out any sprouts by hand. To compost pomace thoroughly, build layers; a layer of pomace, a sprinkling of lime, then a layer of soil, and repeat until the pile is built.

Brush Reducer and Weed-Killer: The high acidity of fresh pomace discourages the growth of many plants. Spread it where you want to keep down weeds and brush.

Wildlife: Deer standing knee-deep in the pomace piles or chipmunks scurrying by, cheeks bulging with apple seeds — something for everyone.

Cider in the Kitchen

Quantities of sweet cider are hard to store for long unless you have a large cooler. This is one reason our ancestors made vast amounts of hard cider. The alcoholic beverage will keep for years if properly made and bottled. You and your family and friends can drink only so much cider, so it's natural that the delicious beverage has an honored place in the kitchen. Sweet cider can be used in pies, baked beans, cakes, sauces, to simmer hams and meats, for jelly, for gelatin desserts. Cider concentrate keeps for a long time and allows you to flavor your cuisine with the unique seasoning of your own cider. Hard cider can be used in any recipe calling for white wine. Both hard and sweet ciders make superior hot mulled drinks. Here are a few of the hundreds of ways to use cider in the kitchen.

NEW ENGLAND CIDER APPLESAUCE

Core and quarter ripe but flavorful tart apples, put them in a stainless steel or enamel pan and cover with fresh sweet cider. Do not use dessert apples. Wild apples are excellent. Cook until the apples are very soft, put through food mill or strain to remove skins, sweeten to taste with maple syrup or sugar, and simmer until the sauce reaches a rich, thick consistency. Pack it into hot, sterilized canning jars, leaving 1/2 inch headroom, and process 10 minutes in a boiling water bath.

APPLE BUTTER

Cook four pounds of quartered (not pared or cored) high acid apples in two cups of fresh sweet cider until soft, about thirty minutes. Work the cooked fruit through a food mill and add sugar (about two cups), cinnamon, cloves, and allspice to your taste. Stir constantly over low heat until sugar dissolves. Cook it down, stirring lest it "catch" and burn on the bottom, until butter is thick enough to spread without running. Pack in hot, sterilized canning jars, leaving a 1/4 inch headroom. Wipe the rims with a clean damp cloth. Seal by taking Mason lid from hot water and placing it on jar, then screwing band on tightly. Process in a boiling water bath for five minutes. Yield is about four pints.

SWEET CIDER JELLO

2 tablespoons gelatin
1/4 cup cold water
1/2 cup hot water
2 cups boiling hot sweet cider
3/4 cup sugar
1/3 cup lemon juice

Dissolve the gelatin in the cold water for five minutes. Add the remaining ingredients in the order given, stirring after each addition. When the sugar is dissolved, pour into a fancy wet mold and chill until firm. Unmold and serve with whipped cream.

BOILED CIDER JELLY

This is cider concentrate carried one step further, and a great favorite in Vermont. Choose ripe, very flavorful apples with a generous proportion of crab apples or tart wild apples. Grind and press the clean fruit on a very cool day — around 40° F. is ideal. Rush the fresh juice to your stainless steel or enamel jelly pan and boil the liquid down rapidly until the jelly "sheets." The jelling point is reached when about seven volumes of fresh cider have been reduced to about one volume of jelly. Slow evaporation over a low, slow fire will result in tough, dark jelly with a caramel flavor. When the jelly sheets, strain it quickly through cheesecloth into waiting hot, sterilized jelly jars. Screw on the lids, then let the jars cool.

BAKED BEANS IN CIDER

This is the wonderful French Canadian way of baking beans. Put four cups of dried beans in cold water to soak overnight. Next morning, put the beans in a kettle, cover them generously with sweet or hard cider, and boil them for half an hour or until the skins split when you blow on them. Put a few strips of pork in the bottom of the bean pot. Roll a whole onion in dry mustard and bury it in the beans. Pour in 1/2 cup maple syrup or molasses. Garnish the top of the beans with a few more strips of salt pork. Add enough cider to cover the beans and bake in a slow over four to six hours. An hour before the beans are done, check to see if they need more cider to keep them moist.

MULLED SWEET CIDER

Put the following spices in a small cheesecloth bag: four or five cloves, a piece of nutmeg, a two-inch piece of cinnamon stick, a cardamom pod. Put the bag in a stainless steel pot and pour over it two quarts of sweet cider. Bring it up slowly to just below the boiling point. Serve hot in mugs.

You can sweeten the mulled cider with honey or maple syrup. Some people like to add a small piece of vanilla bean or a slice of fresh ginger to the spice bag. Keep the spices to a minimum if children are drinking this tasty warmer.

HOT SWEET CIDER EGGNOG

Add six cloves to a quart of sweet cider in a stainless steel pan and boil for three minutes. Add 1/4 cup maple syrup or sugar. Beat six eggs until they are light. Fish the cloves out of the hot cider, and, *very slowly* pour the cider over the eggs, beating vigorously with a wire whisk. Now pour the mixture back and forth from one pre-heated pitcher to another until it is frothy and well mixed. Serve immediately in mugs. Dust the surface of each nog with a grating of nutmeg.

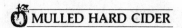 MULLED HARD CIDER

One of the best excuses for winter is so we can enjoy this potent drink which can thaw out a granite tombstone.

2 quarts dry, aromatic hard cider
Strip of fresh lemon peel
1 cinnamon stick
1/2 teaspoon allspice
4-6 whole cloves
1/2 cup maple syrup
1/2 cup light rum

Place all the ingredients in a stainless steel pan and heat over a low fire until the liquid steams. Do not let it boil. Strain it into warmed mugs and serve immediately.